MW01110107

ZOOS

by Emma Bassier

Cody Koala

An Imprint of Pop!
popbooksonline.com

abdobooks.com
Published by Pop!, a division of ABDO, PO Box 398166, Minneapolis, Minnesota 55439.

Printed in the United States of America, North Mankato, Minnesota.

052019
092019

THIS BOOK CONTAINS RECYCLED MATERIALS

Cover Photo: Betty LaRue/Alamy
Interior Photos: Betty LaRue/Alamy, 1; iStockphoto, 5 (top), 5 (bottom left), 5 (bottom right), 9, 11, 13, 17 (top), 17 (bottom left), 19 (bottom left), 19 (bottom right), 21; David R. Frazier Photolibrary, Inc./Alamy, 7; Shutterstock Images, 12, 15, 17 (bottom right), 19 (top)

Editor: Meg Gaertner
Series Designer: Jake Slavik

Library of Congress Control Number: 2018964654

Publisher's Cataloging-in-Publication Data
Names: Bassier, Emma, author.
Title: Zoos / by Emma Bassier.
Description: Minneapolis, Minnesota : Pop!, 2020 | Series: Places in my community | Includes online resources and index.
Identifiers: ISBN 9781532163531 (lib. bdg.) | ISBN 9781532164972 (ebook)
Subjects: LCSH: Zoos--Juvenile literature. | Zoo animals--Juvenile literature. | Visitors to zoos--Juvenile literature.
Classification: DDC 590.7--dc23

Hello! My name is

Cody Koala

Pop open this book and you'll find QR codes like this one, loaded with information, so you can learn even more!

Scan this code* and others like it while you read, or visit the website below to make this book pop.

popbooksonline.com/zoos

*Scanning QR codes requires a web-enabled smart device with a QR code reader app and a camera.

Table of Contents

Chapter 1

Animals Everywhere

People see a dolphin swimming. They watch monkeys climbing trees. They spot an elephant raising its trunk. The people are at the zoo.

Watch a video here!

Chapter 2

A Place to Observe

A zoo is a place to **observe**. Visitors can watch animals eat, play, and sleep. People can learn about animals by watching them.

Learn more here!

Some zoos have many different animals. Other zoos focus on specific animals. For example, **aquariums** focus on fish and other water animals.

Chapter 3

Inside a Zoo

Each type of animal lives in its own space at the zoo. The space is made to be similar to the animal's **habitat** in the wild. Fences or glass walls keep the animal in its space.

Learn more here!

Some animals live inside buildings. Snakes and fish live in glass tanks.

Other animals live outside. Large animals need lots of outdoor space.

People walk from one animal to the next. People read signs about the animals. They learn about what the animals eat. They learn about how the animals act.

The San Diego Zoo in California has more than 4,000 animals!

Das „Who is who" der Zebrawelt

Zookeepers take care of the animals. They feed the animals. They clean the animals' homes. They play with the animals. They keep the animals and people safe.

Some animals play with toys, such as balls.

zoo fence

outdoor habitat

glass tank

Chapter 4

Helping Animals

Many animals are **endangered**. They have a hard time surviving in the wild. They need help. Zoos can give these animals a safe place to live.

Complete an activity here!

Zoos can also teach people about animals. People learn about what animals need. They learn about how to help animals.

Making Connections

Text-to-Self

Have you ever been to the zoo? If yes, what animals did you see? If no, what animals would you want to see at a zoo?

Text-to-Text

Have you read another book about zoos? What did you learn?

Text-to-World

People can observe animals and learn about them at the zoo. Why is it important to learn about animals?

Glossary

aquarium – a building that has many glass tanks filled with water animals for people to look at.

endangered – at risk of dying out.

habitat – the area where an animal normally lives.

observe – to look at and study.

zookeeper – a person who takes care of animals at the zoo.

Index

Online Resources

popbooksonline.com

Thanks for reading this Cody Koala book!

Scan this code* and others like it in this book, or visit the website below to make this book pop!

popbooksonline.com/zoos

*Scanning QR codes requires a web-enabled smart device with a QR code reader app and a camera.